WINDMILLS: In America and Around the World

Text and drawings: James E. Owens

Schiffer Publishing Ltd

4880 Lower Valley Road, Atglen, Pennsylvania 19310

The Eastham windmill was probably built in Plymouth, Massachusetts, around 1680 by Thomas Paine, a millwright from Kent, England. It was moved across Cape Cod Bay in the early 1700s to Truro, in 1793 to Salt Pond in Eastham. In 1808 it was moved to its present site and worked until 1896. The town purchased the mill in 1928. Windmills were often moved from place to place. Arms were the most vulnerable part and often had to be replaced, as you can see here.

Common sails such as the ones on the Eastham windmill were hung by simply climbing up each arm in turn. An elderly Harwich miller failed to secure the brake as he climbed and had to be rescued when a gust of wind began turning the arms as he worked.

In 1710, the Jonathon Young windmill was built in South Orleans, Massachusetts, using timbers from an older mill. It was moved to Orleans Center and used until the 1890s. Captain Hunt floated it to Hyannis Port, Massachusetts, and was given back to Orleans in the 1980s, "flaked" (taken apart), moved, restored by volunteers, and open to visitors in the summer.

The East Mill was built in East Orleans, Massachusetts, in the mid-1700s. In the 1960s, it was restored and moved to Heritage Plantations in Sandwich, where it can be visited today.

The Higgins Farm mill in Brewster, Massachusetts, was built on Ellis Landing Road in 1795. It was given to the Brewster Historical Society, moved to Drummer Boy Park and can be visited in the summer.

1746 saw the construction of the Chatham, Massachusetts, Godfry windmill near Stage Harbor. In 1956 it was given to the town, moved to Chase Park, and can be visited in the summer.

In 1791 Judah Baker built a windmill near Grand Cove in Dennis, Massachusetts. It was moved to South Yarmouth in 1883. Now near the mouth of Bass River, it is restored and open to visitors in the summer.

Ferris Windmill was built in Yarmouth, Massachusetts, in 1732. It was moved to Bass River in 1750, South Yarmouth in 1782, West Yarmouth in 1894, and finally to the Ford Museum in Dearborn, Michigan, in 1835, where it still stands

In 1792 the Nickerson mill was erected in South Harwich, Massachusetts, then moved near Wychmere Harbor. It is part of a private home by Stone Horse Yacht Club and not open to visitors.

Salt was a valuable commodity in ancient times to season or preserve foods and salt works were common on the Atlantic Coast. A typical salt works produced a bushel (65 lbs) of salt per week from 300 gallons of sea water, nine inches deep in a vat exposed to the sun for a week then moved (see chart) to other vats. Movable roofs kept out rain. Left over sludge was processed for Epsom Salts. A Chatham, Massachusetts, owner produced 27,400 bushels of salt in 1837 at $7-8 per bushel. Salt mining replaced the business.

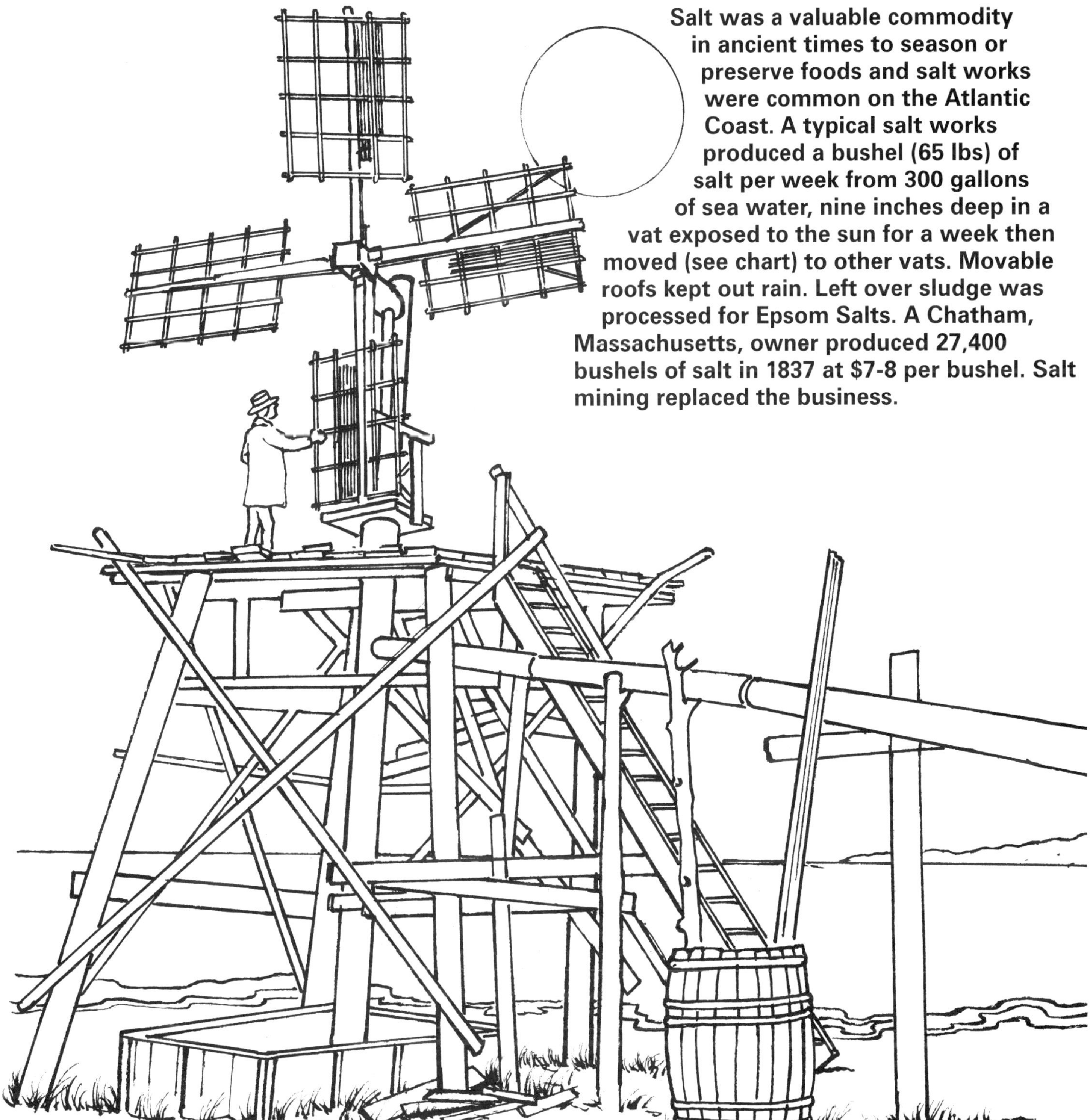

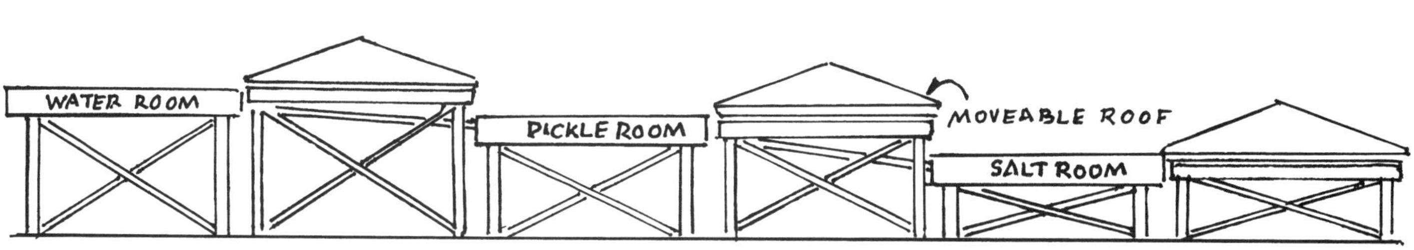

Numerous windmills were built on Nantucket; the second, built by Nathaniel Wilbur in 1746, still exists. In the 1890s, summer resident Caroline French paid $855 for it and gave it to the Historical Society. It is often run in the summertime and is open to visitors.

Block Island, Rhode Island, a separate island off the Rhode Island coast similar to Nantucket and Martha's Vineyard off the Massachusetts coast, had no rivers and relied on the wind to grind flour with many windmills. Unfortunately, none of those Block Island windmills have survived today.

The Jamestown, Rhode Island, windmill was built in1787 on the site of an earlier mill destroyed in the Revolutionary War. It leans slightly to the North due to prevailing westerlies. It is open by appointment through the local Historical Society.

Prescott Farm Mill was built in Warren, Rhode Island, 1813, moved to "Highlands," Fall River, then Quaker Hill in Portsmouth, then Lehigh Hill, eventually to the Middletown-Portsmouth line by Doris Duke's Newport Restoration Foundation and can be visited.

Boyd's Mill stood for many years in Portsmouth, Rhode Island, moved to Middletown, Rhode Island, in the 1990s, and was restored as one of only two existing 8 armed windmills. The other is in England. Some mills had 8, 6, or 5 arms and there was an 8 armed mill in Falmouth, Cape Cod, but it only exists on old postcard today.

Hayground Mill was built in 1809 in Easthampton, Long Island. Used until 1919, but worked only 4 months of the year after 1870, this mill produced 800 bushels of wheat, 400 bushels of corn, and 800 bushels of oats per season.

Pantigo Mill was built in 1771 near Hook Mill, East Hampton, Long Island. It was moved to Pantigo in 1850 and later by John Howard Payne's "Home Sweet Home," where it still stands.

Hook Mill in East Hampton, Long Island, was built by Nathaniel Dominy, a well-known millwright, in 1796. In 1850, a 1/16 share was worth $25. Last used in 1907, it was bought by the town in 1920 and is open to visitors.

The Beebe Mill was built in Sag Harbor, Long Island, in 1820. Because it signaled the arrival of whaling ships, it was known as "Flag in the mill, ship in the bay" mill. It was sold and moved to Bridgehampton in 1837 and last worked in 1911.

In 1799, ship owner James Mitchell built a windmill on Hog's Neck, Southampton, Long Island. He paid 9 pounds, 17 shillings for the "soke" rights or permit to have a mill. In 1812, it went to a miller in Sag Harbor who then sold it to James Corwith for $750. Next it moved to Watermill using 12 yoke of oxen. It ran for 75 years as a "fulling" mill. In 1888, a new summer home cut off the prevailing wind. Damaged in the 1938 hurricane, restored, neglected, and restored anew, it can now be visited.

This unusual wind-powered machine shop owned by Henry Glover of Dedham, Massachusetts, was used in the 1890s and no longer exists, but there were many like it around the USA.

"De Zwann" windmill in Holland, Michigan, was built in the 1760s in the Netherlands. It was purchased, removed to Michigan in 1964 by descendants of Dutch settlers, and put into running order. In the center of a large tulip growing area, it celebrates the heritage of the present population.

This 1848 Danish windmill was purchased by the citizens of Elkhorn, Iowa, mostly of Danish descent, as a bicentennial project. Restored to its former glory as a working mill, it can be visited most of the year.

It took three years to build the Dutch style windmill at Mt. Emblem, Illinois, from 1847 to 1850. Idle since 1917, it is now the focal point of a cemetery there.

Holland's gristmill in Millbank, South Dakota, is over 100 years old and has recently been restored.

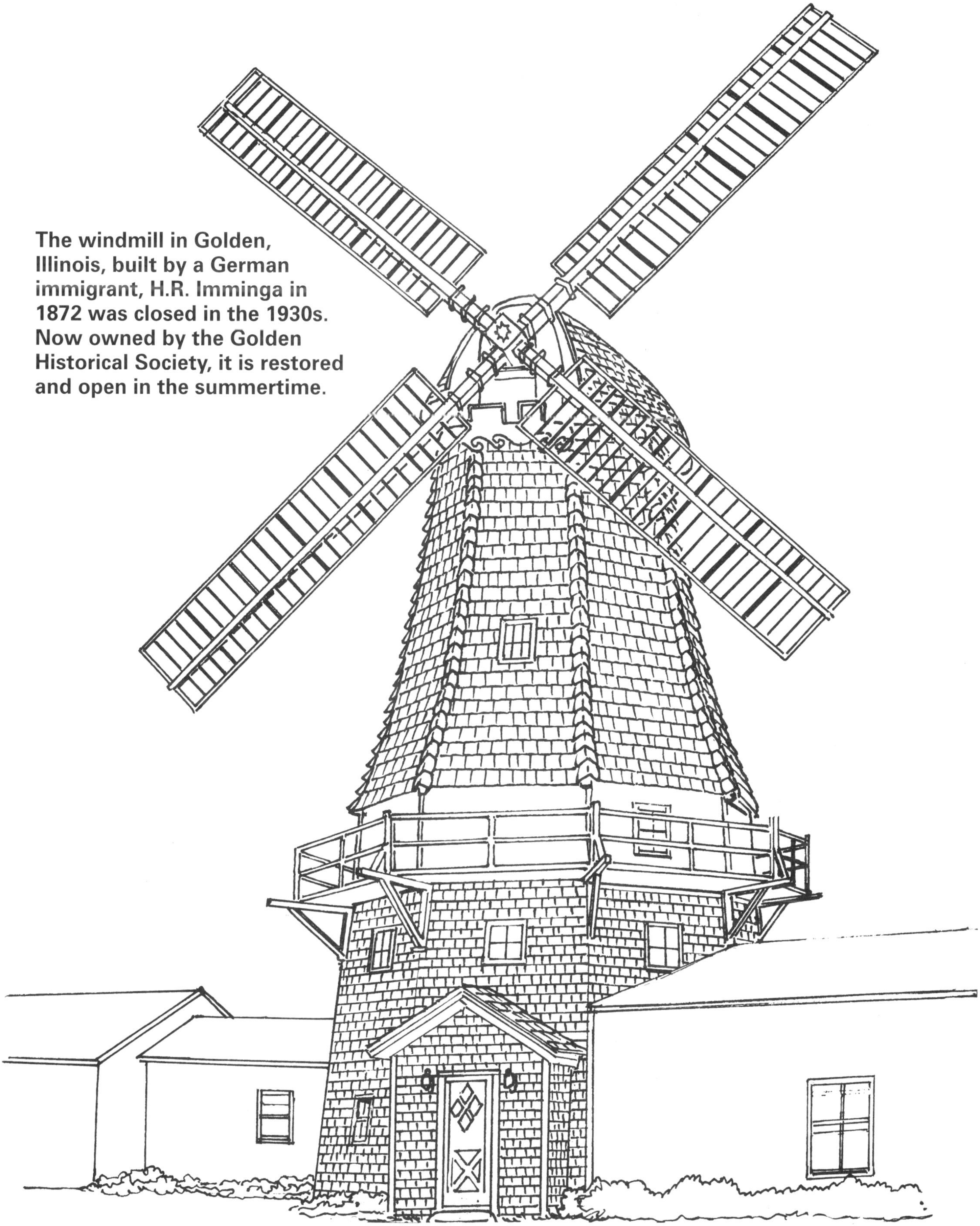

The windmill in Golden, Illinois, built by a German immigrant, H.R. Imminga in 1872 was closed in the 1930s. Now owned by the Golden Historical Society, it is restored and open in the summertime.

De Immigrant windmill in Fulton, Whiteside County, Illinois, is a copy of the Brabant windmill in Vorstenbosch, Netherlands. It was built there and shipped to Fulton, an area settled by people from Friesland and Gronigan. It has buhr stones and is operational.

The Vermeer windmill in Pella, Iowa, was built in pieces in the Netherlands by Luvas Verbij of Hoogemade. Assembled by two Dutch craftsmen in 2002 in Pella, it is representative of the local population and claims to be the tallest windmill in the USA.

In 1877, Peter Burkman built a windmill for Abraham S. Friesen in Manitoba for $2,000. It was torn down in 1920, but Dutch millwright Jan Medendorp in the Mennonite Village in Manitoba, Canada, built a replica. It cost $98,000 to build.

Murphy's Mill (1902) in Golden Gate Park, San Francisco, California, is one of two existing windmills built to water the park. It could pump 40,000 gallons per hour into a manmade lake. The cap and machinery were recently sent to the Netherlands to be restored and the mill is now complete.

This odd windmill was built by German immigrants, Fred Meiss and Otto Fiek, along the Goliad between Indianola and San Antonio, Texas, in 1879. Typically Texan construction of chinked logs, it was moved, after hurricane damage, to Victoria where it was shingled in 1935.

This abandoned windmill stood on Lower Lachine Road near Montreal, Canada. Early French settlers built windmills along the St. Lawrence River in colonial times and some still exist.

Spocott windmill, a post mill, was built in Lloyds near Cambridge, Maryland, in 1878 and ran until a severe blizzard destroyed it in 1888. It was recreated in 1972 to honor US Senator George L. Radcliff, whose family owned the original mill.

This faithful reproduction of Robertson's Mill, the original of which once served colonial Williamsburg, Virginia, now stands in present-day Colonial Williamsburg and has been run in past years.

This post mill is typical of the 65 post mills on Cape Hatteras before the Civil War. None have survived except in old photos. This one is pictured near Kittyhawk.

In 1621 a post mill (the first) was built at Flowerdew Hundred by the James River in Virginia. A 1978 replica was built near the same spot with two pairs (runs) of stones, two spring sails, two canvas sails, and a fantail to "luff" or turn the mill to the wind.

This large post mill is located in Bruge, Belgium, alongside two others and is still a working mill. It is open to visitors.

This Belgian post mill has a housing over its trestle or legs and secondary machinery for processing linseed oil in the lower half as it grinds flour above. The fantail moves the mill into the wind whenever it changes direction.

The "Whip Molen" is typical of small windmills in the Netherlands that pumped out rainwater. Much of the land there is below sea level and flooding is a problem.

This is an unusual version of an early post mill found only in France. The people are wearing costumes typical of Medieval French peasants or millers.

Windmills are somewhat rare in Ireland. This one is the Tacumshane Mill in County Wexford and is probably from the late 1700s. It is built of stone, stucco, and thatch.

The Howe Hill drainage mill in Norfolk, England, one of many to pump rainwater away since much of Norfolk is below sea level. There are many windmills in Norfolk and East Anglia to visit today.

This is the Maude Foster Windmill in Boston, Lincolnshire, England. It has five arms and was built high to catch the wind because of the buildings around it.

Much of the farmland in the Netherlands was recovered from the sea by pumping large areas dry. These "Polders" used many windmills in a row to accomplish the job such as these at Kinderdike. Pumping continues today to get rid of rainwater.

These brick tower mills in Schiedam, Netherlands, were built high because of the many city buildings around them.

Portuguese windmills have "jib" sails and use clay jars on them. The jars whistle and change tone to signal the miller when the wind changes, a unique feature in Portugal.

This Greek windmill is similar to Portuguese mills but with more "jib" sails and no whistling jars. A thatched roof was common in the Mediterranean. Some European mills were also thatched.